40/W.O./3697

A

SEQUENCE

OF

MUSKETRY TRAINING.

(1915, *reprinted, with Amendments*, 1917.)

Issued by the General Staff.

The Naval & Military Press Ltd

Published by
The Naval & Military Press Ltd
5 Riverside, Brambleside, Bellbrook
Industrial Estate, Uckfield, East Sussex,
TN22 1QQ England
Tel: +44 (0) 1825 749494
Fax: +44 (0) 1825 765701
www.naval-military-press.com
www.military-genealogy.com
www.militarymaproom.com

A SEQUENCE OF MUSKETRY TRAINING.

Object.—Training for War.

The Soldier.—To enable him to make the best use of his rifle under all conditions. FIRE DISCIPLINE.

The Commander.—To enable him to make the best use of his men's fire under all conditions. FIRE DIRECTION AND CONTROL.

PAGE.

THE TRAINING OF THE SOLDIER.

(B 9094) Wt. w. 15170—9475   4000   3/17   H & S   P. 17/161 (S)

CARE OF ARMS AND AMMUNITION.

Causes of wear.—(*a*) Friction of bullet; (*b*) heat from explosion of charge; (*c*) pull-through gauze. [M.R., 85, 86.]

Pull-through.—How to pack in butt trap; the three loops; danger of using much worn cord. [M.R., 87.]

Free from grit. [M.R., 89.]

The gauze.—Method of attaching; well oiled; fit tight. [M.R., 89.]

Flannelette.—Regulation only; dry, 4″ × 2″; oiled, use smaller piece. [M.R., 88.]

Oil.—Work well into flannelette. [M.R., 91.]

Result of excessive quantity. [M.R., 88.]

Cleaning of rifle.—Remove bolt; pull through from breech to muzzle; one steady pull; avoid cord wear. [M.R., 88.]

During training boiling water will be used after firing ball. [M.R., 101.]

Bolts.—Clean action and outside; check number of bolt. [M.R., 103.]

Magazine.—Clean inside with dry rag.

Ammunition.—Keep dry and clean; avoid extremes of temperature; never oil the case. [M.R., 116.]

MECHANISM.

(Weights and measurements not required.)

Removing and replacing bolt. [M.R., 94, 95.]
Removing and replacing magazine platform. [M.R., 47.]
Safety catch; bolt lever in lowest position. [M.R., 44.]
Half cock and how to recock. [M.R., 42.]
Bolt-head screwed fully home. [M.R., 46.]

THEORY—I.

References :—*Musketry Regulations, Sections* 17 *to* 30.

1. RIFLING.

Definition.—*See* M.R., Section 19.
History.—Little known of early history ; believed to have been invented in either 14th or 15th century. Reached this country from the Continent about 1594.

2. BEFORE THE BULLET LEAVES THE RIFLE.

(1) *Force of explosion.*—Causes bullet to move from " lead " to muzzle, leaving latter at 2,060 or 2,440 f.p.s., according to whether Mark VI or Mark VII ammunition is used.

(2) *Jump.*—Caused by shock of discharge, and to a certain extent by friction between bullet and rifling.

Jump of two kinds :—

        (*a*) Vertical.
        (*b*) Lateral.

As regards (*a*), whether upward or downward, depends on position of muzzle at moment bullet reaches it. Serious changes caused by altering muzzle velocity.

Vertical jump allowed for by different heights of foresight.

As regards (*b*), caused by lack of symmetry and differences in strength of different parts of rifle. Allowed for by lateral movement of foresight.

(3) *Fixing bayonet.*—Weight of bayonet damps vibrations, hence jump is altered.

(4) *Resting rifle.*—Does not affect jump, therefore shooting unaffected.

(5) *Heated barrel.*—Causes bore to expand, thereby fit of bullet in bore impaired.

(6) *Oily barrel.*—Not of much importance. Causes loss of friction, therefore alters jump.

3. AFTER THE BULLET LEAVES THE RIFLE.

(1) *Resistance of air.*—Checks flight of bullet, *e.g.*, bullet (Mark VII) leaves muzzle travelling at rate of about 800 yards per second. Resistance of air allows it to travel only 600 yards in 1st second, 400 yards in 2nd second, 300 yards in 3rd second.

(2) *Gravity.*—Draws bullet downwards with ever-increasing velocity. Thus path or trajectory of bullet is curved instead of straight.

(3) *Elevation and sighting.*—Necessary to allow for fall of bullet due to gravity, by directing line of departure as much above target as bullet would fall below it. Sights provided to enable this to be done, and at same time keep target in view.

(4) *Drift.*—*See* M.R., para: 156.

(5) *Dangerous space.*—*See* M.R., Section 26.

(6) *Angles of descent.*—Necessary to know, since they affect the dangerous space. Individual fire to great extent limited by size of dangerous space.

One minute equals 1 inch for every 100 yards. *See* Trajectory Table. The last figure in Trajectory Table shows how much the fall is in the last 100 yards.

$$\text{Gradient of descent of a bullet} = \frac{\text{fall in feet}}{300.}$$

4. GENERAL.

Ricochets.—*See* M.R., para. 168.

Firing up or down hill.—*See* M.R., Section 28.

Atmospheric conditions.—*See* M.R., Sections 29 and 30.

5. SOUNDS CAUSED BY RIFLE FIRE.

Two sounds are made when a rifle is fired—the noises of the explosion and the noises of the bullet passing through the air. When the bullet is going faster than sound, the latter noises reach the ear of an observer who is in front of the rifle, as a single sound or crack which is much louder than the noise of the explosion. When bullets are fired across the front of an observer this noise appears to come from a point well down the course of a bullet, and is therefore no guide as to the position of the firer.

VISUAL TRAINING.

NECESSITY FOR INVISIBILITY AND HOW OBTAINED.

Necessity on account of :—Accuracy of modern weapons.
Obtained by :—(i) Close study and use of ground. [I.T., 108.]
          (ii) Suitable formations. [I.T., 90 and 118.]
          (iii) Night training. [I.T., 113.]
          (iv) Neutral tinted uniforms.
          (v) Smokeless powder.

TRAINED EYESIGHT NECESSARY OWING TO INVISIBILITY OF ENEMY.

When—(i) on battlefield ; (ii) working alone. [M.R., 306 ; I.T., 110.]

DIFFICULTIES TO BE OVERCOME.

    (i) Differences in sight: town and country.
    (ii) Brain power not developed.
    (iii) Lack of words—MILITARY VOCABULARY. [M.R., 310.]

STANDARD TO AIM AT.

"That of a stalker."

1. Ability to distinguish enemy from surroundings.
2.   „   „ aim accurately at service marks.
3.   „   „ report on what seen.
4.   „   „ recognise objects described.
5. Establishment of feeling of self-reliance.
6. Study and use of ground.

USE OF FIELD GLASSES.

Usually to confirm what has been seen with the naked eye, but may be used to search for special targets. [M.R., 309.]

A SYSTEM OF INSTRUCTION.

For the private soldier :
Instruction must begin early and be progressive. [M.R., 307.]

BARRACKS.

Lectures. Men taught to recognise their immediate surroundings. Military vocabulary. [M.R., 308.]
Training on landscape targets or Solano targets. [M.R., 280.]

7

Open Country.

Object.—To teach men to locate low service targets up to 800 yards. [M.R., 306.]

(*a*) *Silhouette targets,* of different sizes, shapes, and colours, arranged against various backgrounds. Limits of areas in which targets are, clearly marked. Targets counted. Their characteristics and positions described. Reasons for difference in their visibility brought out. [M.R., 308.]

(*b*) *Fatiguemen.*—Instead of targets. Movement quickly detected. Blank used to train ear to locate sound. [M.R., 308.]

(*c*) *Practice.*—Two squads assume service positions and try to locate each other. [M.R., 309.]

Test of elementary training.—Locating four fatiguemen up to 800 yards. [M.R., 299 (i).]

Examination of Ground.

Object.—To enable men to make clear reports, to understand instructions, and to recognise features of military importance. [M.R., 309 ; I.T., 110.]

(*a*) *Definite line* in landscape. Described in detail.

(*b*) *Areas of ground.*—Clearly-defined boundaries. Description of general shape. Natural and artificial features. Trees and fences. Fields, &c. Features of military importance brought to notice by questions. Military vocabulary largely increased.

As progress made.—Squad examines ground and gives description from behind cover. Limited time given for examination by squad; then squad turned about to give description. Large areas divided into sections: foreground, middle distance, and background.

(*c*) *Road work.*—Cultivate an eye for country by making men observe what they pass on the march. Question them after a given interval.

Test of elementary training.—[M.R., 297 (ii).]

Recognition.

Object.—To train the firer to recognise targets described. *Recognition* means a soldier's understanding of the exact point at which his commander wishes him to aim. [M.R., 306, 308.]

Practice in recognition.—Officer or N.C.O. describes aiming point. Must be an expert. Men lay rifles on point recognised (aiming at ground).

On landscape targets squad turns about, and each man separately shows *exact* point of aim with a stick.

Test of elementary training.—Each man should recognise four points. [M.R., 299 (ii).]

JUDGING DISTANCE BY EYE.

Recruits, up to 600 yards. [M.R., 314.]

Trained soldiers, up to 800 yards, up to which distance they should not make a mean error of more than 100 yards. [M.R., 304, 323.]

Officers, non-commissioned officers, and selected men up to 1,400 yards. Beyond that distance, judging is very inaccurate and instruments are probably available. [M.R., 303, 324.]

The limit of individual fire (600 yards) must be recognised by all. [M.R., 314.]

The MEAN of the estimates of several individuals is generally more reliable than the estimate of one individual.

A SEQUENCE OF INSTRUCTION.

Unit of measure.—Some familiar distance is used as a unit; generally necessary for men to be taught one; 100 yards is a convenient unit; the class place themselves independently at what they think is 100 yards from an object; the distance between the farthest and nearest is paced; 100 yards is measured accurately from the object; the class is shown the correct unit. This method can only be used when the whole of the ground to be measured is visible; the unit is applied to ground between the class and flags; fatiguemen are not used, otherwise this method may be confused with others; examples are shown to which this method cannot be, or can only be partly, applied. [M.R., 312.]

Appearance.—The appearance of men in different positions, and of objects of known size, is studied and noted, at various distances and under all conditions of light, background, atmosphere, &c. [M.R., 312, 314.]

The following points should be noted :—

(1) The apparent height of the object.

The foresight, or other guide, may be used for comparing the apparent height of objects at different distances. [M.R., 318.]

(2) Appearance of the heads and shoulders of men.
(3) Distinctness of outline.
(4) Distinctness of the face, hands, rifle, and head-dress.
(5) Movements when loading and firing.

Appearance varies with eyesight of individuals; a classification range is suitable for the early lessons; the system is applied to men, then to objects of known size.

Opportunities for revising the impressions of the appearance of men at various distances, should sometimes be given. [M.R., 319.]

Objects are over-estimated when they are difficult to see or when the eye is attracted by other objects. Objects are under-estimated when they are clearly distinguishable. [M.R., 315.]

Comparison with known ranges and similar methods [M.R., 317, 318]:—

(1) This includes trying to halve the distance, and judging the half-distance first.

(2) Judging to some object of known size, and then getting the distance by comparison.

(3) Judging with the assistance of range cards.

Bracketing.—Decide on the longest distance the object can be; decide on the shortest distance the object can be; take the mean. [M.R., 317.]

Practice.—Constant practice is necessary under all circumstances, both in peace and war, as serious errors must be expected under strange conditions. [M.R., 305.]

Methods should be combined till distances can be approximately judged by the general impression conveyed to the eye. [M.R., 313.]

Time limits should be gradually introduced after the first lessons. [M.R., 318.]

To prevent guessing, reasons should be given by everyone for their estimates. [M.R., 316.]

Lying in the open or suitable positions behind cover, should be used normally when judging distance. [M.R., 325 (ii).]

By constant practice, mean errors should be reduced to 10 per cent., but an average error of 15 per cent. must be expected; records of results should be kept. [M.R., 305.]

Test of elementary training. —[M.R., 299 (iii).]

Quarterly tests.—[M.R., 322 *et seq.*]

AIMING INSTRUCTION.

The sights of all rifles must always be in perfect order. [M.R., 200.]

System of instruction.—Illustration; explanation; imitation; examination. [I.T., 10 (2).]

When illustrating, free use to be made of diagrams on paper or blackboard, practical illustration with the rifle, use of most suitable appliances, *e.g.*, accuracy of aim—the Le Gret aim teacher; aiming off for movement—the aim corrector.

Preliminary arrangements.—Pre-arranged programme useful to help non-commissioned officers to carry out instruction.

Length of lesson.—Should be short, 20 to 30 minutes. [I.T., 4 (6), 10 (3).]

Example of hour's work at beginning of training :—

First 20 minutes	... ...	Trigger pressing.
Second „	... ...	Accuracy of aim.
Third „	... ...	Care of arms.

Individual instruction necessary. [I.T., 10 (1).]

Progressive sequence of instruction is necessary. [I.T., 10 (4).]

Instruction to be combined with visual training and judging distance in the later stages.

A SEQUENCE IN AIMING INSTRUCTION. [M.R., 198.]

1ST OR ELEMENTARY STAGE.

Object.—ABSOLUTE ACCURACY. Aim laid from an aiming rest and tripod, at a special target, not over 100 yards. [M.R., 204, 201.] [M.R., Pt. II., Plate 29.]

SIGHTS.

Reasons for sights. [M.R., 152.]
Graduations. [M.R., 30.]
Method of adjusting. [M.R., 31.]
Practice.
Check. [M.R., 202.]

ACCURACY OF AIM.

Method of using sights (3 rules) diagrams. [M.R., 202.]
Correct aim shown. Le Gret. [M.R., 203.]
Practice.
Common faults explained as they occur, shown by Le Gret or
paper; backward men shown correct aim in stages with
Le Gret. [M.R., 202, 206.]

Reasons for aiming at centre of lowest part of mark.—Whole
mark is kept in view; better chance of hitting a vanishing
target; better chance of hitting if distance over estimated;
counteracts tendency to shoot high; (assists close grouping in
collective fire).

Reasons for full sight.—Less tendency to vary amount of foresight; facilitates rapid aim.

Test of elementary training for accuracy and consistency. Triangle
of error. [M.R., 207–209, 299 (v).]

AIMING OFF FOR WIND. (Small bullseye at about 10 yards.)

Necessity for; the service method; method of keeping
elevation, diagrams; aim laid showing method (no measurements); practice. [M.R., 173, 214.]
Description of various winds. (Mild, Fresh, Strong.)
[M.R., 213.]

2ND OR ADVANCED STAGE.

Object.—To retain ACCURACY at more difficult targets, aiming
rest and tripod used where possible. Figure targets up to 600
yards to ensure focussing. [M.R., 201.]

ACCURACY OF AIM.

Correct aim shown; practice; check.

ELEVATION TABLE.

Range.	Elevation.	Vertical rise. Mark VI.	Mark VII.
200	300	1 foot	6 inches.
300	400	2 feet	12 „
400	500	3 „	20 „
500	600	4 „	30 „
600	700	5 „	42 „

Use of table taught with full size targets, spotting disc and
cardboard sights.

12

AIMING OFF FOR WIND.

Effect of various winds on bullet at 500 and 300 yards; units
of 1 foot shown at various distances up to 600 yards. [M.R.,
638 (A).]
Practice in aiming off, judging feet at actual distances.
[M.R., 215.]
Judging allowance for various winds. [M.R., 214.]
Correct point of aim shown by fatiguemen with discs. [M.R.,
216.] In aiming off all distances are taken from centre of lowest
part of mark; a target's breadth will be measured in a similar
manner.

ACCURACY AND RAPIDITY OF AIM.

Object.—To quicken the aim whilst retaining absolute
accuracy; taught with the aiming disc; time saved (1) in coming
to aiming position, (2) in quickness of aim; quickness must not
be obtained at expense of accurate trigger release; practice.
Test of elementary training with aiming disc. [M.R., 299 (viii).]

AIMING OFF FOR MOVEMENT.

Necessity for. [M.R., 222.]
Mechanical movement of rifle taught under 100; aim
corrector; practice. [M.R., 223.]
Table of allowances for movement. [M.R., 224.]

3RD OR FINAL STAGE.

Object.—To retain ACCURACY and QUICKNESS, at service targets
at all ranges. [M.R., 201.]

ACCURACY OF AIM.

At silhouette figures, or men in service positions up to 600.
[M.R., 210.]

AIMING OFF FOR WIND.

At silhouette figures, or men in service positions up to 600.
[M.R., 214.]
Test of elementary training with aiming rest at men.
[M.R., 299 (vii).]

AIMING AT GROUND.

At natural objects up to 600 yards.
Marking down an enemy; preliminary arrangements with
fatiguemen necessary to ensure their re-appearance at exactly
the same place. [M.R., 210.]

ACCURACY AND RAPIDITY OF AIM.

Snapshooting at vanishing targets at actual distances; early
lessons short distance, exposures long and in same place each
time; later lessons up to 300, shorter exposures, targets appear
in different place each time.

Necessity for :

(1) Constant watch of front.
(2) Quick aim.
(3) Instant re-loading.

AIMING OFF FOR MOVEMENT.

Snapshooting at moving targets at actual distances, correct allowances.

Progressive lessons as for accuracy and rapidity of aim [M.R., 225.]

Aim corrector used to check at slow moving targets.

Test of elementary training with aim corrector. [M.R., 299 (vii).]

AIMING UP AND DOWN.

Object.—To avoid necessity of making petty alterations of sights. [M.R., 217, 221.]

Limited to three feet above or below regulation point of aim. [M.R., 226.]

Practised at fatiguemen or vanishing targets at various ranges. [M.R., 218.]

AIMING AT GROUND over 600 (recognition), including long range sights.

Aiming at natural objects described verbally. [M.R., 211, 205.]

Test of elementary training with aiming rest. [M.R., 299 (ii).]

AIMING OFF FOR WIND over 600 (recognition), including long range sights.

Men practised in carrying out instructions accurately. [M.R., 215, 205.]

Necessary for Officers and N.C.O.'s to know wind table. [M.R., 638 (A).]

Point of aim indicated :—

(1) With reference to breadth of target.
(2) With reference to intervals in a formation.
(3) By use of an auxiliary aiming point.
(4) (If above impossible) in yards or feet.

Fatigueman shows correct point of aim. [M.R., 216.]

LONG RANGE SIGHTS.

Use of these taught on the above principles as far as they apply. [M.R., 205.]

Reasons for use :—

(a) Less strain on neck muscles than high backsight elevations.
(b) Greater rapidity of aim at indistinct objects.
(c) Less tiring to eye, as only foresight and aiming mark have to be focussed. Eye looks *through* aperture, but *at* open backsight.

A SEQUENCE OF INSTRUCTION, LONG RANGE SIGHTS, worked in
with former sequence.

1st Stage :—
Sights ; accuracy of aim.

3rd Stage :—
Aiming at ground over 600 ; aiming off for wind over 600.

FIRING INSTRUCTION AND MUSCLE EXERCISES.

The press-off of all rifles should always be in perfect order. [M.R., 108.]

Rifles must fit the firers. [M.R., 119.]

Individual instruction necessary; small squads; normal positions suit nearly all men. [M.R., 229; I.T. 4 (4).]

Squads formed in semi-circle for the purpose of explanation; for practising the firing positions this formation may be retained or a straight line may be adopted, in both cases great care must be taken that all men are facing so that their lines of fire are parallel, sufficient targets must be provided to enable this to be carried out. [M.R., 228.]

Method.—Illustration; explanation; imitation; examination. [I.T., 10 (2).]

Firing rest may be used if necessary. [M.R., 231.]

Dummy cartridges must always be used, but must be carefully inspected. [M.R., 262; I.T., 4 (8).]

Firing instruction should proceed simultaneously with aiming instruction. [M.R., 199.]

Progressive sequence of instruction essential. Sequence must fit in with stages reached in other branches of instruction, *e.g.*, men not allowed to snap till they have learnt correct aim, trigger pressing, correct firing position. [I.T., 10 (4).]

Targets used should be similar to those used in Aiming Instruction according to progress.

A SEQUENCE IN FIRING INSTRUCTION.

TRIGGER PRESSING. [M.R., 232–237.]

Importance.

Forefinger capable of independent movement; grip of three fingers and thumb; use of and position of forefinger on trigger; direction of pressure; double pressure required.

First pressure taken as butt is brought into the shoulder [M.R., 250]; second when aim is correct; breathing restrained during second pressure.

Recruit sitting, elbows rested on table; rifle rested on sand-bags; butt not to be in shoulder.

Instructor's hand over recruit's; recruit's hand over instructor's; concentration of mind necessary; practice.

POSITIONS IN OPEN.

STANDING.

Targets pointed out ; complete firing position shown.
When used. [M.R., 240–241.]

(1) Loading position shown.
　Method of reaching position; practice; faults
　corrected. [I.T. 10 (5).] (Sequence of checking:
　if feet wrong, everything else is, so start at feet.)
　Reasons for half-turn, &c.

(2) Method of loading shown.
　Imitated. [M.R., 242.]
　Method of unloading shown. Practice; faults
　corrected. [M.R., 243.]

(3) Method of adjusting sights shown. [M.R., 244.]
　Practice. [M.R., 219, 202.]
　Sights checked, faults corrected.
　Test of elementary training for sight setting. [M.R.,
　299 (iv).]

(4) Aiming position shown. [M.R., 249.]
　Method of reaching position; practice; faults
　checked, sequence of checking. [M.R., 250.]
　Reasons for movements.
　Test of elementary training for firing positions. [M.R.,
　298 (i).]

(5) Completed firing positions shown.
　Point of aim declared ; practice. [M.R., 198, 235,
　252.]
　Test of elementary training for trigger pressing with
　aim corrector or sub-target. [M.R., 236, 299 (vi).]

MUSCLE EXERCISES.

Muscle exercises should proceed simultaneously with aiming
and firing instruction. [M.R., 199.]

Object; conspicuous target; rifle always approximately
aligned and first pressure taken; practice—daily by recruits,
constant by trained men ; all positions. [M.R., 266.]

POSITIONS IN OPEN (*continued*).

The following are taught and tested on the same principles as
the standing position :—

LYING. [M.R., 253–255.]

KNEELING. [M.R., 256 (ii).]

SITTING. On ground on which this position would be used.
[M.R., 256 (i).]

LYING POSITION adapted to show small target surface.
[M.R., 255.]

POSITIONS BEHIND COVER.

Eyes must be kept on target between shots when firing from cover. [I.T., 108 (5).]

ARTIFICIAL COVER.

All positions adapted to various forms of cover ; taught and tested as in the open ; squad views from front ; correct uses of cover explained ; each kind of cover used discussed. [M.R., 257 ; I.T., 108.]

NATURAL COVER.

Use of folds in the ground and cover from view demonstrated ; practice in using suitable positions on broken ground. [M.R., 259.]

RAPID LOADING.

Taught in all positions. [M.R., 261.]
Test of elementary training.—[M.R., 299 (ix).]

RAPID FIRING.

(Not to be taught till a man can fire deliberately in all positions.)

Rapidity in loading and aim combined ; accuracy necessary ; rifle loaded in shoulder. [M.R., 260.]
Test of elementary training.—[M.R., 299 (x).]

AUTOMATIC ALIGNMENT

is the involuntary action of the muscles in carrying out correctly what they have constantly practised ; consists of bringing the rifle to the shoulder and aligning the sights before pressing the trigger ; constant practice and correction of faults in peace essential for this to be done correctly in war. [M.R., 197.]

FIRE DISCIPLINE TRAINING.

References—I.T., Sec. 117 ; *M.R., Sec.* 55.

A form of training which ensures that men obey orders rapidly and accurately, and, when left to themselves, use their rifles to the best tactical advantage.

After sufficient progress in aiming and firing instruction, practice will be given in moving in extended order.   [M.R., 267 ; I.T., 90.]

Preliminary collective exercises consist of simple practices to teach accurate and quick obedience to fire orders, and to practise quick concentration of fire on various targets. Divided into three stages.   [M.R., 276.]

Two first stages of this training are simply drill, no tactical considerations. Third stage, lessons more advanced. [I.T., 90 (5).]

Normal firing position is lying ; this will always be used unless other orders are given. The instructor will always make certain that the aiming mark he describes is visible to every member of his squad.   [M.R., 285 ; I.T., 92 (4).]

Standing, kneeling and sitting positions will only be practised under conditions when they would be used, or, in wet weather the position used may be standing or kneeling, when this will be explained beforehand, and such position will be used without further orders.   [M.R., 286.]

Rapid fire should never be ordered or allowed unless the target justifies its use.

An assumed position of the enemy, always to be pointed out, to which men turn when halted. The position of the instructor will be where he can best supervise the work of his squad, and he must constantly check and correct faults. [I.T., 92 (3) (6).]

1st Stage.

Easy aiming marks used. Squad halted at ease, extended in line to one or two paces. The instructor gives orders for loading. He then gives the range, and the men adjust their sights. The fire orders are then completed, and the men act on them. They will continue to fire until the order to " Cease fire " or " Unload " is given, or until the named number of rounds have been fired. [M.R., 285.]

The passing of orders along the line should also be practised. The main points for the instructor to note are :—

(*a*) Position assumed by the firer, dexterity in manipulation of bolt, loading, safety catch, and pouch buttoned.

(*b*) Correct adjustment of sights.

19

(c) Recognition of the target.
(d) Difference between "Rapid" and "Deliberate" fire.
(e) Difference between "Cease fire" and "Unload."
(f) Re-charging of magazine.
(g) Alertness of the men in attending to fresh orders.
(h) Passing of orders. [M.R., 288.]

Every irregularity must be checked.

2ND STAGE.

Easy aiming marks used. Rifle loaded before the exercise begins; the men are responsible for keeping their magazines charged. Movement introduced. The squad advances in an extended line.

On the command or signal "Halt," the squad assumes the lying position (unless otherwise ordered). Fire orders are given (particular attention being paid to correct pauses so that each part of the order may be acted on before the next is begun).

The instructor will walk round his squad, paying attention to the points mentioned under "1st Stage."

Passing of orders will also be practised. [I.T., 96 (2).]

3RD STAGE.

Movement; use of ground and cover; initiative and judgment. Squad advances in an extended line.

A fatigueman appears (for about a minute). Instructor orders, "At the fatigueman—Fire." Squad halts, each man assumes the position he thinks suitable, adjusts his sights, and fires. [M.R., 290, 291.]

The instructor will first take up his correct position as commander, and then act as in "2nd Stage." [I.T., 92 (8).] As proficiency increases, more difficult targets used, e.g., men or carts passing on road, &c. Change of targets, aiming off, &c. [M.R., 291.]

Anticipatory orders may sometimes be given in this stage. [M.R., 292.]

Test of elementary training.—[M.R., 298 (ii).]

Some of the duties of the soldier in fire discipline :—

(1) Recharge his magazine on every possible opportunity.
(2) Make proper use of the safety catch.
(3) In advance, to get up and down quickly. [I.T., 92 (5).]
(4) When advancing select his next halting place, and move straight to it.
(5) Make best use of cover.
(6) Never press the trigger unless his sights are aligned on the mark. [I.T., 116 (12).]
(7) Observe the enemy.

In Collective Fire.

(8) Adjust sights for range ordered.
(9) Recognise aiming point described.
(10) Count the number of rounds fired if necessary.

[M.R., 273.]

(11) Limit his rate of fire to that ordered.
(12) Pass orders.   [M.R., 288.]

⌊*In Individual Fire.*   [I.T., 116 (2), 123 (12–14).⌉

(13) Carry on the fight.
(14) Select targets.
(15) Judge distance.
(16) Adjust sights.
(17) Alter point of aim from observation.
(18) Use rate of fire necessary.
(19) When possible join nearest commander.
(20) If wounded place ammunition where it will be found.

21

MINIATURE RANGE.

Miniature ranges should be ·WARM, WIDE, LIGHT, and WELL EQUIPPED. Normal type fulfils all those conditions.

Range.—25 yards. [M.R., II., 122.]

Floor.—Asphalt, concrete, wood or ashes ; smooth, so as to prevent dangerous ricochets. [M.R., II., 109.]

Bullet Catchers.—Simplest form, a steel plate, sloped so as to deflect bullets into sawdust. [M.R., II., 133.]

Firing Point should be raised.

Marking.—Best system by use of field glasses, targets being brought to firing point at end of practice.

Range Discipline.—*Rifles should always be laid down with the bolts open when anyone is in front of the firing point.* [M.R., II., 90 ; M.R., I., 367.] When rifles are being loaded, unloaded, or inspected, they should be directed towards the target.

Heating.—Stove near firing point most economical.

Light.—Skylights over firing point and targets. Artificial light; electric best, but acetylene or coal gas satisfactory. A row of burners giving a total of 250 candle power necessary ; two positions, one on ground for landscape target practice, electric only, the other 3 ft. 6 in. from ground for ordinary target practice. [M.R., II., 127, 128, 136.]

Equipment—

Cover of all kinds can be made with sandbags ; trenches should be made where possible.

Rifles should be service rifles bored for miniature ammunition, so that firers may become accustomed to the weight, length, balance, bolt action and sighting of the rifle he would use in war, otherwise miniature shooting cannot be a satisfactory preparation for service shooting. [M.R., II., 85.]

Rifles must be cleaned after every 20 rounds. [M.R., II., 88 ; I., 366, 105.]

WAR OFFICE MINIATURE RIFLE : a toy. [M.R., II., 89 ; I., 52.]

SERVICE RIFLE, solid bore ·22 ; latest pattern has floating striker. [M.R., II., 86 ; I., 49.] Correct sighting for direct hits, 300. [M.R., I., 360 ; 49 (12).] PARKER RIFLING.

Ammunition.—Only miniature ammunition may be used on a miniature range.

Target apparatus should be suitable for :—

(a) Range practices.
(b) Individual field practices.
(c) Collective field practices.

(A) *Hythe Pattern*, useful for outdoor ranges. Sawdust or turf banks, former very good for observation practices.

(B) *Solano Pattern*, 10 feet, Mark I. Cost £9 5s. 0d., best pattern.

Accessories drawn to scale for use at 25 yards, still difficulties of service shooting can only be partly reproduced, *e.g.*, difficulty of estimating range; effect of wind; effect of atmosphere on bullet; effect of atmosphere on eyesight; shock of discharge.

Training.—A certain standard of training necessary before shooting begins.

Range Practices.—With or without a rest or cover, various positions.

Grouping.—Rings 1, 2, and 3 inches, or S.R. and Territorial Force, 1 inch larger. Recruits must reach a 3-inch standard.

Test of elementary training.—[M.R., I., 299 (xi), 356.]

Application.—Wind gauge used to represent necessity for wind. First at bullseye targets, then at figure targets. [M.R., 359, 356.]

Snapshooting.—First at figure targets, later at silhouette targets. [M.R., 363.]

Crossing Targets, wind gauge used to necessitate the correct allowance for movement. [M.R., II., 197 ; M.R., I., 363.]

Rapid Fire.—Magazine experimental.

Individual Field Practices.—Most individual field practices can be fired using Solano figures representing men up to 600, and then target apparatus with scenic accessories.

Collective Field Practices.—The necessity for collective fire can be shown and many useful field practices carried out on the Solano target, or on landscape targets.

Landscape Targets.—The frame for landscape targets is 10 ft. long and 5 ft. high. Landscape pictures in sheets, 5 ft. by 2 ft., are pasted on to the lower portion, leaving 3 ft. of blank sky-screen above to receive the shots. [M.R., II., 156.]

Many of the landscapes are more than 2 ft. high, and must be cut down, as this necessitates very high backsight elevation, and affects the safety of the range. Frequent change of landscape targets is desirable, as the features become well known. When firing at landscape targets the rifles should be given extra elevation so that the bullets will strike the blank sky-screen, even if aim is taken at an object at the bottom of the landscape. So that all the rifles should hit at the same height above the aiming point they should be harmonised. A board should be hung in every miniature range, showing the elevation required for shooting at landscape targets. [M.R., II., 157.]

30-YARDS RANGE.

Very useful at all times, especially when classification ranges are distant for training recruits, and for indifferent shots throughout the year, with service ammunition. [M.R., 369; II., 46.]

Desirable that each corps should have one. [M.R., II., 48.]

Lateral protection greater than normal danger area of a classification range. [M.R., II., 54.]

Load with rifles pointing towards targets, otherwise a shot clearing the top of stop butt would go at least 2,500 yards. [M.R., 369.]

Suitable for all practices which can be fired on miniature ranges. [M.R., II., 47.]

Advantages over miniature ranges :—

Man uses his own service rifle of which he knows the pull off. [M.R., 370.]

Learns to shoot with it under easy conditions. [M.R., 369.]

Becomes accustomed to the shock of discharge, and any tendency to flinch is eliminated. [M.R., 369 ; II., 46.]

Becomes accustomed to noise of discharge which is greater than on an open range. [M.R., II., 51.]

Rapid fire with service cartridges can be used. Practice in loading with service ammunition. [M.R., II., 46.]

Practice with long-range sights is possible. [M.R., 370.]

RANGE PRACTICES.

STAGES IN MUSKETRY TRAINING OF SOLDIERS.

(1) Elementary training. [M.R., 343, 350.]
(2) Training on miniature and thirty-yards ranges. [M.R., 369.]
(3) RANGE PRACTICES. [M.R., 5.]
(4) Individual field practices.
(5) Collective field practices. [M.R., 516.]

Range practices are an advanced stage of elementary training, and must be regarded as such. Object, to ensure that a certain standard has been reached by recruits, and kept up by trained soldiers, before they go on to more practical shooting. They are only fired up to 600 yards, the limit of individual fire. [M.R., 415.]
The individual musketry training of the soldier not complete till he has become a good service shot. Range practices are simple as compared with service shooting; in some respects directly opposed to service methods. [M.R., 591.]

STANDARD.

Standard required in range practices has been advanced within last few years. Formerly men could be marksmen without any practice; now constant practice in handling the rifle is necessary. Standard in elementary work ensured by tests of elementary training. Range practices only a waste of ammunition if this standard not kept up. [M.R., 351, 238, 296.]

WHEN FIRED.

Recruits do not begin range practices till they are ready; trained soldiers must be always ready. [M.R., 344, 372, 351.]
Range practices, especially those for recruits and classification, should be fired under favourable weather conditions. [M.R., 437, 427.]

FIRING POSITIONS AND USE OF COVER.

Regulation positions are obligatory in all range practices except those fired from behind cover, when the positions must be adapted to the cover. Cover must not be specially prepared as a rest. [M.R., 391, 462, 457, 449, 256, (2) 455.]

BASIS OF PRACTICES.

A course of range practices is based on a system of progressive instruction, and every practice is framed to illustrate some tactical use of fire, or some essential point of elementary training.

GROUPING.

Grouping means firing a series of shots, usually 5, at a distinct aiming mark without any alteration of sighting or point of aim. Not fired at distances over 100 yards. The diagram made by the shots is called a group. [M.R., 378, 443, 381.]

Grouping brings out the necessity for :—

Absolute accuracy and consistency of aim, correct holding and trigger pressing, and control of the nerves.

Men should see their groups measured and discussed.

The method by which the cause of a bad group is discovered is called the "analysis of faults." [M.R., 384.]

Analysis of Faults. [M.R., 379.]

Rifle is tested by an expert shot to show the soldier that the rifle is not to blame, or to discover if the rifle is inaccurate.

Aim tested by the triangle of error method.

Trigger pressing, tested with aim corrector.

Sight tested, near, by reading; distant, by counting distant objects. [M.R., 382.]

Nerves the probable cause, if the above are correct. [M.R., 383.]

When a man has made a bad group, his faults should be analysed at once before leaving the range. A note should be made on the register of remedies to be used. [M.R., 384, 443.]

If Rifle is at Fault.

A soldier should not be expected to make considerable allowances to counteract the error of his rifle. [M.R. 123.]

He must fire with his own rifle. [M.R., 458.]

Throw of the rifle should be corrected by the armourer. [M.R., 130, 131.]

APPLICATION.

Application practices follow grouping; these teach the firer to adjust his sights and point of aim, so as to apply his shots to a mark. Application brings out the necessity for knowledge of the wind and elevation tables, confidence in powers of shooting, and ability to aim for the next shot according to the point of aim at the moment of firing and the result of the last shot. [M.R., 386.]

A man must be able to group before he can hope to apply with confidence. [M.R., 386, 444.]

As correction in sighting is seldom possible in individual firing in war, it is most important to estimate the elevation and point of aim for the first shot. [M.R., 418, 419.]

Time limit for each shot in slow practices—20 seconds. [M.R., 445.]

SNAPSHOOTING.

Snapshooting means firing an effective shot in the shortest possible time. These practices bring out the necessity for watching the front, quickness of aim, observation of the strike of the bullet, change of point of aim from observation, and immediate reloading. [M.R., 394, 395, 447, 449.]

RAPID FIRE.

Rapid fire means firing as many rounds as possible with accuracy in a given time. Rapid fire brings out the need for clean and quick loading, and handling of arms, quickness of aim and working the bolt with the rifle in the shoulder. These practices give men an opportunity of finding their best rate. [M.R., 394, 447, 450, 449, 397; I.T., 116 (12).]

When rapid and slow practices are fired at the same distance in classification, each man may fire the rapid immediately after the deliberate practice. [M.R., 438.]

Range practices for the Regular Army and Special Reserve begin with qualifying practices; if the standards are not reached preliminary training has failed in its object. [M.R., 373.]

RECRUITS' COURSE.

Divided into four parts. Part I. Fired at short ranges during elementary instruction, and is to give confidence; it may be repeated as often as necessary, and practices may be altered to suit individual requirements. The wind gauge is permitted so that recruit may not have to aim off the centre of target too far, as he has quite enough to think about otherwise. The use of a rest is allowed to help the recruit to steady his rifle; the practices are then repeated without a rest. Differences between a rest and cover. [M.R., 374, 401.]

Part II. The conditions are more difficult, and the figure target is introduced. A certain standard must be reached; if not the practices are repeated. [M.R., 402.]

The coloured figure target prevents men dwelling on their aim, taking a fine sight, or focussing the foresight instead of the target at the moment of firing. [M.R., 393.]

Part III. Still more advanced. Rapid fire, snapshooting, and the use of cover is introduced.

Part IV. Is a test of progress. So that recruits may regard range practices as elementary they fire some simple field practices at the end of their course. [M.R., 403.]

TRAINED SOLDIERS' COURSE.

Divided into three parts. Part I. Is a test. Those who do not qualify repeat before going on to classification practices, after having fired certain practices (7–14) of Part II. This is to show the importance of maintaining a standard in elementary shooting. [M.R., 425, 426, 398.]

Part II. Instructional. The best instruction must be given. The number of rounds and order of practices can be varied by officers commanding companies. [M.R., 417, 398, 619 (xi).]

Conditions may be varied by commanding officers. [M.R., 417.]

Advisable for rounds to be saved from good shots in slow practices. [M.R., 394.]

In instructional practices the wind gauge may be used to enable a man to aim at some part of the target, otherwise its use is not desirable ; the fine adjustment should not be used. [M.R., 387.]

Part III. Is a test in which the use of the wind gauge and fine adjustment, or any instruction, is absolutely forbidden. Sights may be altered at any time in classification practices, except as above. [M.R., 452, 451.]

Allotment of Ammunition.

Surplus ammunition. [M.R., 421, 139, 422.]

FIRING-POINT INSTRUCTION.

1. An instructor can only watch one man at a time, so men now fire singly. [M.R., 461.]

2. The man must be watched, not the target, or faults cannot be discovered. [M.R., 376, 392, 390, 391.]

3. No hurry should be allowed; it is better to discuss the reasons for failure of a few shots properly than to hurry over many. [M.R., 442.]

4. A true declaration of the actual point of aim at the moment the rifle was fired must be made before the shot is signalled, so that true deductions can be made. [M.R., 390.]

5. Neither the firer nor his rifle should be touched. [I.T., 10 (5).]

6. The firer should not be told anything; if necessary he should be made to reason out the causes and remedy for failure by questions.

FIELD PRACTICES.

In field practices targets must be looked upon as an actual enemy, and service conditions must be observed. [M.R., 550.]

The effect of the enemy's fire and the nervous tension which it causes is absent, therefore results are better than can be expected in war. [M.R., 560, 553.]

Quick opening of fire and effect from the first shot is essential. [M.R., 419, 548, 509, 523, 564.]

Targets which fall when hit add interest. [M.R., 547, 549.]

The object of the practice and the special lessons to be brought out should be explained beforehand. [M.R., 527, 553.]

No interference (except for safety) during the practice. [M.R., 530, 550.]

Full criticism of good and bad points on completion. [M.R., 527, 563.]

Ammunition in excess of that to be used may be issued to practice men in looking after it and in recharging the magazine whenever possible.

INDIVIDUAL FIELD PRACTICES.

Individual skill produces decisive effect at close ranges.

In range practices the soldier has applied practically some of the lessons of elementary training. He should have confidence in his shooting and know the peculiarities of his rifle at known ranges up to this limit of individual fire. [M.R., 503.]

Conditions have been comparatively easy, and he must apply his knowledge in firing at difficult service targets at unknown range. [M.R., 506, 507.]

Practical use of the following additional elementary points brought in :—

1. Use of ground.
2. Location of low service targets.
3. Choice of targets.
4. Judging distance.
5. Quick opening of fire.
6. Application from observation (partly practised in snap-shooting).
7. Choice of rate.
8. Mutual assistance. [M.R., 509, 529.]

High scores in range practices bear no relation to results of firing under service conditions, even in peace time. [M.R., 504, 419.]

Distance, not over 600. Practices progressive as regards targets, distance, &c. [M.R., 529, 543.]

Separate target for each man. [M.R., 524.]

Men fire in pairs; one fires, the other observes. Observer should not use field glasses. In later practices men may fire singly to bring out necessity for self-reliance. [M.R., 524, 526.]

Firers of each detail should be regarded as a squad. [M.R., 530.]

Markers are used to signal hits. [M.R., 526.]

Men of a squad should mutually assist each other. [M.R., 524, 526.]

Key ranges, over 600, may be given sometimes to bring out their use. [M.R., 528.]

Movement of firers and targets must be included. [M.R., 524.]

Skill in snapshooting and rapid firing must be increased. [M.R., 508, 525.]

Practice necessary in snapshooting standing during a *rapid advance.* [M.R., 515.]

Points for criticism :—

Loading at first safe opportunity; use of safety catch; selection of point from which to fire; method of advance to fire position; use of cover on position; watch on front; consultations as to range of objects useful as range marks; consultation as to wind; on targets appearing, consultation as to range and point of aim; quickness in opening fire; instant reloading; methods employed in firing at a target advancing; quickness in gaining effect; points of elementary training, both firer and observer; observer's power of observation and of reporting strike of bullet; method of correction for next shot; information given to rest of squad; rate of fire used; recharging of magazine.

COLLECTIVE FIELD PRACTICES.

Beyond 600 fire effect can only be assured by means of collective fire. [M.R., 505, 269 ; I.T., 116 (7) (iii).]

Collective field practices primarily intended to give all commanders practice in their duties of fire direction and control. [M.R., 523, 544, 542, 540, 510.]

Also gives practical experience in use of fire and its effects on various targets. Practice to men in applying what they have learnt in fire discipline. [M.R., 542, 510.]

Fire direction, control, and discipline often neglected at training with blank ammunition, therefore ball ammunition must be used to show their importance. [M.R., 511, 541 ; I.T., 107 (10).]

Collective field practices should be fired over 600 yards. [M.R., 543.]

Programme depends on ammunition available, time, and range facilities. All ammunition allotted for field practices must be fired in these. [M.R., 540.]

Most instruction with least ammunition ensured by a battalion programme of tactical demonstrations. [M.R., 540.]

Progressive training of all ranks. In early stages for small fire units, simple tactical schemes illustrating a situation which may be expected on service. Schemes should be simple and give separate instruction in each phase of a combat. [M.R., 548.]

The scheme as a rule will illustrate either attack or defence. [M.R., 552.]

All possible arrangements should be made behind cover, or in anticipation. [M.R., 568.]

Mutual support should be frequently practised both with and without control. [M.R., 551.]

Company exercises should be fired when the units composing them and their commanders have proved sufficient skill. [M.R., 568.]

Range finders should be used and key ranges made during preliminary reconnaissance for use from fire positions to be occupied later. [M.R., 568.]

Combined field firing should be practised when ground admits. [M.R., 569, 571.]

Points for criticism :—

All points for fire direction, control and discipline—these points are all enumerated in "Fire Action II" and "Fire Discipline Training."

DEMONSTRATIONS.

These may be fired when the points which they bring out best fit into the general programme. [M.R., 554.] *e.g.*:

Penetration tests during elementary training in the use of cover. Demonstration of results of faults in aiming during aiming instruction. Demonstration of use of ground during training in fire discipline. Demonstration of rapid and accurate fire before range practices. [M.R., 555.]

Demonstration showing limit of individual fire before individual field practices. [M.R., 531.]

Demonstration of grouping of concentrated fire, and of degrees of distribution during collective field practices. [M.R., 556, 557.]

Test of vulnerability of formations, use of ground and cover, &c., during collective field practices. [M.R., 561, 562.]

RECORDS.

Records of collective field practices and demonstrations should be kept for reference. [M.R., 519, 520, 559, 560, 558, 565, 566, 567.]

VISUAL TRAINING—II.

A SYSTÉM OF INSTRUCTION.

For the commander.
The system followed by the private soldier, then the following:

INDICATION.

Indication means the shortest and most easily understood description of an aiming point by a commander.

Indication and recognition necessary to ensure close grouping of collective fire. 75 per cent. of shots probably wasted owing to bad indication and recognition. [M.R., 532, 533.]

One system advisable in a battalion, or better, throughout the Army. *Aiming points* must be described as seen with the naked eye. *A front* always to be pointed out. [M.R., 278.]

Taught in stages.

1st Stage.—Description of aiming points, without aids.

2nd Stage.—Description of different objects, using aids. [M.R., 279.]

Aids only used when absolutely necessary.

(a) *Reference points.*—Prominent objects, or others of military importance. Two hand-breadths apart. Reasonably distant. Of different kinds. Names by which they are known must be made clear to all.

(b) *Finger breadth method.*—Shows roughly the distance of an object from a reference point. Only left hand should be used. Arm must be held straight out. Fingers vertical unless object is immediately above or below reference point; then horizontal. Reference point and object both kept in view. Necessarily inaccurate. As progress made, fingers only used as a check Trained men should seldom hold up fingers.

(c) *Clock ray method.*—Shows the direction of an object from a reference point. Clock-face vertical; centre on the reference point; not used on level ground when object, reference point, and observer, are on same level. Direction right or left should be given as well as clock ray.

(d) *Combined method.*—(b) and (c) together; seldom necessary.

Practice in indication.—Key rifle laid on point to be indicated, or pin showing point on miniature landscape target, or point on landscape target actually pointed to. Officer or N.C.O. indicates point. Squad recognise it.

Test.

Unofficial.—Indication of four points. 80 per cent. of squad must recognise each. Squads must be reliable. Practice in indication during winter. Necessary to be expert before company training

THEORY—II.

1. This lecture, which deals with collective fire, has as its object the explanation of Sections 31 to 34, Musketry Regulations, Part I.

All officers and N.C.O.'s must have a working knowledge of above.

2. Plate XII, M.R., represents a large target fired at by an individual, without alteration of sighting elevation or point of aim. Note the following points :—

(1) All the shots are not in the same place.
(2) The shot holes are more numerous in the centre.
(3) Approximately half the shots are above the centre horizontal line, the other half below.
(4) Approximately half the shots are on the right of the centre vertical line, the other half left.
(5) The distance from the topmost shot and the lowest one is greater than that between the extreme right and left shots.

From above we deduce the following :—

(1) Since shots are not in same place, it follows that the trajectories of bullets do not coincide. The figure thus formed is known as the cone of fire.
(2) Since shot holes are more numerous in the centre we know that the cone of fire is denser in the centre than on the outside.
(3), (4) and (5) show us that the cone of fire is not circular but oblong in section, and that its density decreases uniformly from the centre to the outside.

3. Plate XIV, M.R., represents the size of rectangles which will contain the shots fired by an expert under the most favourable conditions, at a service target at different distances.

From this will be seen what a small chance even an expert has of hitting a prone figure at distances over about 600 yards. Compare the chances of the average shot firing, say, on a windy day, when he is tired or hungry.

4. Since we cannot expect fire to be effective over 600 yards when employed individually, we must use collective fire. Good illustration is the manual fire engine :—

The men working the engine represent the men of fire unit.
Water produced by their efforts represents stream of bullets.
Man holding the nozzle represents the fire unit commander.

Unless everyone works in conjunction no results can be hoped for.

5. Cone of fire from a number of rifles is larger than that from one, since skill varies, eyesight, &c. The size will be still further increased if firers are tired, aiming mark is hard to see, &c., &c.

6. Imagining that collective fire has been applied to a large vertical target marked with two concentric rings, and that a long mat has been laid out behind this target so as to collect all bullets passing through it:

That portion of the mat struck by all the shots passing through the target is known as THE BEATEN ZONE.

That portion of mat struck by shots passing through the centre ring is known as THE NUCLEUS OF THE BEATEN ZONE.

That portion struck by bullets passing through both the centre and larger ring is known as THE ZONE OF EFFECTIVE FIRE or EFFECTIVE BEATEN ZONE.

THE NUCLEUS contains 50 per cent. of shots fired.

THE EFFECTIVE BEATEN ZONE contains 75 per cent. of shots fired.

THE BEATEN ZONE 90 per cent. of shots fired.

The remaining 10 per cent. are too far out to be included.

7. Useful results can only be expected if the target is included within the EFFECTIVE BEATEN ZONE for any range.

Experiments have shown that as range increases the size of the EFFECTIVE BEATEN ZONE (E.B.Z.) decreases. This is due to increased angle of descent of bullet. Beyond 1,500 yards it increases again, especially laterally, owing to increased effects of errors in aiming, &c., &c.

Under favourable peace conditions it has been found that the size of the E.B.Z. varies very little when fired by different units.

Sizes of E.B.Z. on level ground are as follows :—

Mark VI.: at 500 yards—220 yards; 1,000 yards—120 yards; 1,500 yards—100 yards ;

Mark VII.: 500 yards—330 yards ; 1,000 yards—180 yards; 1,500 yards—150 yards.

Lateral measurements same for Mark VI. as Mark VII., as follows :—

500 yards, 7 feet ; 1,000 yards, 14 feet; 1,500 yards, 28 feet.

8. The permissible error in ranging is equal to half the depth of the E.B.Z. for any particular range, *e.g.* :—

Assume target to be 1,000 yards distant.

E.B.Z. for 1,000 Mark VII. is 180 yards.

If range is obtained absolutely correct, half E.B.Z. will be one side of target, half the other.

If error of over 90 yards is made (*i.e.*, half E.B.Z.), whole of E.B.Z. will miss target.

9. The PROBABLE ERROR in ranging will sometimes be greater than the permissible error. The probable error may be taken as follows :—

1. When using instruments ... 5 per cent. of range.
2. Judging distance ... 15 „ „
3. 1 and 2 combined ... 10 „ „

Therefore sometimes necessary to increase depth of the E.B.Z. Only advisable if the following conditions are complied with :—

(1) *Distance not under* 1,000 *yards,* under 1,000 yards E.B.Z. sufficiently deep to counteract probable error.
(2) Fire unit contains at least 100 firers, if less than 100 the density of the E.B.Z. will be so reduced as to render the fire insignificant.
(3) *Observation of the strike of bullets impossible.*
(4) *Surprise effect required.*
(5) *Men of fire unit unshaken.*

The method by which the E.B.Z. is increased in depth, which is an indifferent substitute for accuracy in ranging, is known as COMBINED SIGHTS, or SEARCHING.

Example to illustrate procedure :—

Distance to target judged to be 1,000 yards.
Probable error at 1,000 yards (15 per cent.) is 150 yards.
Target may therefore be anywhere between 1,150 and 850 yards.
Ground which must be searched is 300 *yards.*
The E.B.Z. for 1,000 yards (Mark VII) is 180 yards.
Amount of ground left unsearched will be 120 *yards.*

Put sights of half the fire unit to 1,050 yards, and the other half to 950, which is a difference in sighting elevation of 100 yards.

We now get the following :—

Two E.B.Z's. which overlap, viz. :—
E.B.Z. for sights at 1,050 yards extends from 1,140 yards to 960 yards.
E.B.Z. for sights at 950 yards extends from 1,040 yards to 860 yards.
Ground searched is from 1,140 yards to 860 yards, which is 280 yards, the bullets being more numerous where the overlap occurs.
This leaves a total of 20 yards unsearched using Combined Sights as against 120 yards when using one elevation.

Note.—It is impossible to try and search this 20 yards by increasing the difference between elevations, since if this is done a gap of 20 yards will be left in the centre of the combined E.B.Z., which is the spot where the target is most likely to be. From this we get the rule.

Rule.—*Never use more than* 100 *yards difference between elevations when using Combined Sights.*

(B 9094)   D 2

10. Effect of slope of ground on the size of E.B.Z.

Most important for officers and N.C.O's. to realise effect produced by E.B.Z. falling on to ground which is not horizontal, since it will guide them in use of ground. Cover, and Formations (dealt with in separate lecture).

Musketry Regulations, Part I, paras. 185 to 191, and Plates XVII and XVIII.

RANGING.

Ranging.—The means adopted for ascertaining the sighting elevation required to hit an object.

Range and *sighting elevation* are not always the same: when firing up or down hill; when barometric pressure and temperature are not normal; when a head or rear wind is blowing. [M.R., 169, 170, *et seq.*]

Necessity for ranging.—Errors in elevation cause greater loss of fire effect than personal errors in shooting at distances over 600 yards. If sighting is incorrect, the least effect will be obtained when shooting is best. [M.R., 302, 301; I.T., 116 (15).]

Principal methods of ranging :—

(1) *Judging distance by eye.*—Most frequently used; has been fully dealt with. Average error about 15 per cent.
(2) *Observation* of the strike of bullets, or of their effect on the enemy. [M.R., 301.]

The best means, but not always possible. [M.R., 330; I.T., 116 (16).]

Sufficient volume of fire must be employed to ensure observation. [M.R., 331.]

Short bursts of rapid fire often simplify observation. [M.R., 331; I.T., 116 (12).]

The fire must be closely grouped. [M.R., 331, 196.]

Objects in vicinity of the target may be used to range on, for better observation. [M.R., 333.]

One elevation well under the estimated range should first be used. Beware of the few short or over shots. Bold alteration of sighting necessary. [M.R., 332; I.T., 116 (16).]

The best position for observation, above and behind the firers. [M.R., 334.]

(3) *Range-taking with instruments.*—Special courses now held to ensure greater efficiency; inseparable from fire action; fully and practically trained range-takers invaluable; badly-trained and inaccurate range-takers dangerous; choice of range-takers, lance-sergeants or corporals; good and well-trained eyesight; strong; intelligent; good writing; good military vocabulary; able to write short, clear, descriptions; self-reliant; good map reading.

Training of Range Takers.

(A) *Technical*, by a specially selected officer, includes :—Ranging under all conditions at service objects (greatest error should not exceed 5 per cent.) ;.testing the instrument daily and carrying out certain adjustments. Description of objects :—As seen with the naked eye, when acting as an observer, or when describing range marks ; and as seen through the instrument, when describing objects which need not be recognised till they are approached nearer ; the method of preparing range cards.

(B) *Tactical*, by the company commander, includes :—All that is necessary to enable the range-taker to use the above qualifications to the best advantage on active service ; a sound tactical knowledge necessary to enable him to forestall the company commander in his requirements as regards ranges. The company commander alone can decide on the position of his range-takers. The following is a guide :—

(1) *On the march.*—He should be near his commander ; as a rule at the head of the company. Should the company commander be sent on in advance of the battalion to reconnoitre, &c., range-taker should accompany him when possible. [I.T., 98 (7).]

(2) *At preliminary reconnaissance* prior to attack.—Whilst the officer commanding battalion is issuing his orders for attack to the company and machine-gun commanders, their range-takers should attend, when possible. Company commanders may then give their range-takers such orders as are necessary to guide them in the preparation of range cards. Range cards prepared whilst company, platoon, and section commanders are reconnoitring. Several copies necessary. [I.T., 122 (4) ; M.R., 301.]

(3) *During the attack.*—Constant check on the distances already taken necessary. When enemy's exact position located, new range cards made out if possible, new ranges taken, as required by company commander, or as judged necessary by range-taker. Instrument useful as a telescope for observation, Marindin 17 diameters, Barr & Stroud 14, Position of range-taker during attack, firing line when ranges are wanted, but seldom suitable ; supports more freedom of action, but difficult to pass ranges to firing line ; generally confined to frontage of company ; local reserves generally too far in rear ; near company commander probably best. [I.T., 116 (15).]

(4) *After assault.*—After a successful assault, range-taker should prepare range cards to assist in the pursuit of the enemy with fire against counter-attack, and act as an observer ; should his company be detailed for the further advance, he should prepare the necessary range cards. [I.T., 124 (6).]

(5) *Rearguard action.*—Range cards from each position occupied ; can be taken to next fire position before retirement ; range cards to be used over line of retirement, should occasion demand ; very similar to those used in advance.

(6) *With outposts.*—Range cards from all positions held by picquets and to be held by supports; all range-takers may be used; probably useful with one of picquets by day as observer; by night with supports. [I.T., 129 (3).]

(7) *Defence.*—Range cards made out for all trenches to be held by company. [I.T., 129 (3), 116 (9).]

(8) *During peace training.*—Officers commanding companies, from a knowledge of the range-takers, should know what instructions are necessary; careless range-takers are apt to give away positions; slackness must be watched for; generally passed over in peace owing to want of confidence in the instrument due to inefficient range-takers; value of accurate ranges cannot be shown except during collective field practices; combination of instruments and judging distance, 10 per cent. of error.

Other Methods of Ranging. [M.R., 301.]

Use of maps.—Generally too-small scale; difficult to locate exact positions. [M.R., 319.]

Sound.—Count at rate of eleven beats in three seconds; each beat equals 100 yards.

Information from other troops.

Forward or back reckoning, after having obtained a range.

RANGE CARDS.

Preparation of range cards for attack.—Should always be used in attack; first objective is a line or point which is to be made good in the advance; need not be exact position of enemy; if first objective is over 2,000 yards, range-taker must either approach to within that distance or take preliminary key ranges. [I.T., 122 (4).]

Ranges taken in *direct* line of advance.

0	Description of first objective	2000 Very accurate
	(Unmistakable objects, described so that they will be recognised when reached.)	
500	Row of fir trees	1500
1000	Hedge with pollarded trees	1000
1400	Wind pump...	600

Preparation of range cards for defence.—(*a*) Mark off on card position from which ranges are taken; (*b*) describe position accurately; (*c*) select an unmistakable object and draw a *thick* setting ray to it; (*d*) draw two semi-circles representing the 600 and 1,000 yards limits; (*e*) select objects to range on— "range marks"—these should be positions, &c., which enemy will occupy or have to pass, or obstacles, *e.g.*, bridge, gap in thick hedge, barbed-wire fence, &c.; (*f*) draw ray as accurately as possible to show direction of object, and of a length corresponding to the distance; keep card set; (*g*) write short descriptions (or draw representations) of the objects at the end of the ray; (*h*) write the distance to each object under the description; (*i*) avoid too many rays, which are apt to become confusing; (*j*) when possible, make one ray do for more than one object. [I.T., 102, 129 (3).]

All men to be made familiar with distances; pass to relieving troops. [I.T., 129 (3).]

LATERAL JUDGING DISTANCE. [M.R., 318.]

All officers, non-commissioned officers and scouts, should know some measurement which will cover laterally one-tenth of a forward distance; measurement can be obtained by covering 10 yards at 100 yards, then applying at longer distances.

GROUND. COVER. FORMATIONS.

Close connection between musketry and manœuvre. [I.T., 1 (10); 108 (1).]

Necessary to close with enemy to gain decisive victory. [I.T., 90 (2).]

All variations of combat, either attack or defence; impossible to lay down fixed system of either; general principles exist, these must be applied with common sense, according to ground and situation. [I.T., 114, 118 (1).]

Cover of two kinds.—From fire and view; folds in ground most common form. [I.T., 108.]

Other types :—

Advantages of cover : Concealment during advance or whilst firing ; a few men well concealed may check a great number ; protection from fire gives confidence ; rest for rifle.

Disadvantages of cover: Cover from view alone useful when enemy does not know it is occupied, otherwise attracts fire ; men lying in the open with a suitable background and avoiding all unnecessary movements often form a very difficult target; may prevent the free use of the rifle ; even cover from fire may form a good aiming mark ; may be a range mark; may delay the advance ; may cause crowding. [I.T., 108.]

May attract men out of direct line of advance and expose them to oblique fire. [I.T., 123 (11), 139 (5).]

Isolated cover.

Cover from aircraft. [I.T., 108 (7) (8), 118 (11).]

ATTACK.

Advance assisted by :—

    (a) Reconnaissance before and during attack. [I.T., 121, 122 (6).]

    (b) Use of ground, cover, and formations. [M.R., 185.]

        (1) To produce fire effect. [I.T., 103.]

        (2) To reduce loss.

    (c) Fire action of all arms to assist movement. (Separate lecture.)

Preliminary reconnaissance by regimental officers and N.C.O's. [I.T., 122 (4) ; I.T., 123 (3).]

As thorough as time permits.

Objectives pointed out.

Ranges taken ; lines of advance within frontage selected and reconnoitred ; object, to advance as far as possible *unseen* ; precautions by officers and scouts; stray shells and unaimed fire must be expected. [I.T., 121 (4).]

Large areas probably covered by unaimed fire. [I.T., 118.]

Very small columns often suitable ; cover difficult owing to steep descent. [M.R., 190.]

Under frontal artillery or direct long-range infantry fire.—Very small columns often suitable ; on irregular front ; the distances and intervals are to a certain extent decided by the forward and lateral effect of shell fire ; small columns facilitate use of ground and control ; this formation should be kept as long as possible. [I.T., 118 (3), (4).]

If surprised by artillery fire, best to continue advance. [I.T., 118.]

Within effective infantry range.—On open ground extend ; steady rapid advance best. [I.T., 118.]

Successive lines moving simultaneously, usually not less than 200 yards distance. [M.R., 183.]

On forward slopes.—Depth of beaten zone is lessened as slope increases ; shallow formations ; successive lines may be closer. [M.R., 187, 188.]

Often best to make one rush down a long slope. [I.T., 121 (12).]

In wooded country.—Successive lines may be closer.

On reverse slopes.—Depth of beaten zone greatest when slope and trajectory are parallel ; unaimed fire as above. [M.R., 189.]

Firing line on a crest.—On a very steep reverse slope ; supports and reserves close up at all ranges. [M.R., 191.]

On a gentle reverse slope, supports far in rear when enemy at long range ; close up in defiladed zone when enemy at short range. [I.T., 123 (7).]

Distances settled by company commander or other officer and continually varying throughout the attack. [I.T., 123 (7).]

When fire effect is required.—Necessary to deploy ; extension depends on volume required ; fire effect chief consideration. [I.T., 118 (5).]

Fire positions should afford :—Free use of the rifle ; sufficient field of fire ; concealment ; background ; cover from fire ; a rest for the rifle.

When checked by enemy's fire.—Advance by rushes. [I.T., 121 (12).]

Considerations before advance.—Reconnaisance ; strength of advance ; the ground ; method of advance ; distance to be covered ; a question of ground and condition of the troops. [I.T., 121, 123, 108 (2).]

Mutual support. [I.T., 123 (6).]

Rushes sudden and simultaneous ; fire seldom opened between fire positions ; advantage must be taken of unexpected bursts of covering fire from artillery, machine-guns, or specially detailed bodies of infantry, or friendly cavalry charges, &c. [I.T., 121 (16), 118 (8), 92 (5).]

At close infantry range.—All ground practically swept by fire ; under heavy fire rushes necessarily shorter ; covering fire must be heavy ; parts of line on favourable ground work forward ; creeping, crawling, &c., exceptional ; units must be reformed under cover. [I.T., 93 (11), 121 (12).]

Sometimes necessary to entrench in attack. [I.T., 121 (13), (9).]
Under machine-gun fire.—If caught in close formation take
cover; extend on wide front if possible; use of cover, ground,
and background; irregular advance by small groups; avoid
ground favouring observation of fire; avoid aiming marks.
Against cavalry.—Protect flanks; steady timely fire; any
formation suitable for effective fire; dead ground favours cavalry.
[I.T., 118 (7).]

DEFENCE.

Active defence only can be decisive. [I.T., 134, 126 (3).]
Siting of trenches in defence.

(1) Choice of ground.
(2) Clearance of foreground, 400 yards. [M.F.E., sec. 1,
28, 29.]
(3) Concealment.
(4) Cover, lowest possible parapet.
(5) Creation of obstacles.
(6) Communications.

1. General position chosen for tactical reasons. Line to be
held chosen to give maximum fire effect with least exposure to
artillery fire.
Field of fire may be often sacrificed to latter. [I.T., 127 (1).]
Ground over which enemy must advance to be covered by
direct or flanking fire or both. [I.T., 129 (1), (4), 116 (11).]
2. A clear field of fire, of at least 400 yards if possible.
[M.F.E., 29; I.T., 129 (3).]
3. Avoid hill-tops and prominent salients. High-sited
trenches facilitate reinforcing, &c., but often leave dead ground.
(I.T., 129 (4), 140 (4).]
Fire plunging. [M.R., 167.]
Low-sited trenches facilitate concealment, grazing fire. Often
impossible to reinforce. [I.T., 129 (4).]
4. Fire trenches. Parapet cannot be too low, if field of fire
sufficient; eye must be brought to level of firer's eye when
siting; must assimilate background; *sharp lines to be avoided.*
[M.F.E., 29.]
Must be bullet proof; $3\frac{1}{2}$ to 4 feet of earth at top; 5 to
6 feet of sodden earth or clay; loopholes; parados. [M.F.E., 2.]
Cover for supports. [I.T., 127 (3).]
5. Break up enemy's formations; force enemy to take most
exposed lines of advance; delay enemy under close fire.
[M.F.E., 42; I.T., 129 (5).]
6. Necessary for free movement of defenders in rear of line
held.
Allow counter attack. [I.T., 127 (4), 129 (5).]
Trenches between supports and firing line. [I.T., 129 (6).]

FIRE ACTION—I.

Fire alone can seldom win a battle. Necessary to close with enemy to gain a decisive victory. Impossible to lay down fixed system of attack or defence. General principles exist, must be applied with common sense. [I.T., 90 (2).]

To produce best fire effect, best use must be made of the ground occupied by both oneself and enemy, and of the knowledge of vulnerability of targets. [I.T., 118 (1); M.R., 185.]

ATTACK.

Principal uses of fire in attack :—

To assist advancing troops to get close to enemy. [I.T., 121 (6).]
To prepare the way for assault.

Fire seldom opened in attack when satisfactory progress can be made without it. [I.T., 116 (8).]

When advance is checked, fire opened by parts of the firing line which cannot advance, or by covering troops to enable front line to reach good fire positions. [I.T., 123 (5), 116 (8).]

Company commander, when in a position to do so, decides when fire is to be opened, subject to orders from his battalion commander. [I.T., 116 (6).]

Considerations. [I.T., 116 (7), (8), 121 (5)] :—

(a) The supply of ammunition.
(b) Beyond 1,400 yards infantry fire seldom of much real value.
(c) Probability of effect.
(d) Prevention of heavy loss.
(e) Position of troops given way.
(f) Advance delayed.

Target, normally that which is checking the advance. Sometimes vulnerable targets, e.g., machine guns in their travelling carriages, artillery on the move, infantry in close formations. [I.T., 116 (7), 118 (9).]

DEFENCE.

Supply of ammunition being easier :—Fire may be opened at longer range than in attack to produce premature extension with loss of control and delay. Generally best for decisive results to wait. [I.T., 116 (9), (13).]

COLLECTIVE FIRE.

The fire of any number of rifles used under the orders of a leader is called Collective Fire. Collective fire is used as long as possible [M.R., 270, 505, 273, 176] :—

(a) Keeps men in hand.
(b) Checks the expenditure of ammunition.
(c) Enables fire to be used to the best tactical advantage. [M.R., 505.]
(d) Gives best assurance of correct sighting and point of aim.
(e) Only means of surprise by fire.
(f) Gives best results.

Means of controlling collective fire :—

(a) Naming number of rounds. [M.R., 273.]
(b) Passing of orders along firing line. [I.T., 96.]
(c) Signals. [I.T., 94.]
(d) Short whistle. [I.T., 95.]

Collective fire may be concentrated or distributed. [M.R., 271.]
Concentrated fire is the fire with one point of aim of all the rifles under a leader.
Used : Against very vulnerable targets. [M.R., 271.]
To produce great effect at a particular point. [I.T., 116 (10).]
Ranging. [M.R., 331, 196.]
Partial distribution.—The fire of two or more groups of rifles under a leader, each group using a separate point of aim. Used when effect is required at several points at the same time.
Maximum distribution.—The fire of all the rifles under a leader, each rifle using a separate point of aim on a given front. Used when some effect is required all along a part of the enemy's line, *e.g.*, to prevent movement, or, to disturb the enemy's aim, to cover movement or entrenchment. [I.T., 116 (10).]
Combined sights.—Fire with combined sights is the fire of two groups of rifles using one aiming point, but two elevations. [M.R., 195.]
To obtain fire effect the target must be in the effective beaten zone. [M.R., 192, 181.]
On service errors of ranging (except by observation) at distances over 1,000 yards may be so great as to prevent fire effect. To increase the chance of effect half the rifles may be sighted 50 yards over and half 50 yards under the estimated range; the beaten zone is thus made deeper.
Used : By not less than 100 men. [I.T., 116 (16).]
When some effect must be obtained immediately. [M.R., 196.]
When observation of fire has failed, and serious errors in ranging are expected.
When the target cannot be exactly located.
May be used in defence against an advancing enemy by reducing the higher elevation by 200 yards at a time.

INDIVIDUAL FIRE.

Individual fire is fire opened without orders from a leader. Never used when fire orders can be given. Seldom justified over 600 yards. [M.R., 268.] Control must be assumed as soon as possible. [M.R., 272.]

RATES OF FIRE.

Deliberate fire.—The normal rate, five to six rounds a minute. [I.T., 116 (12); M.R., 274.]

Rapid fire.—The best rate of individuals. A reserve of power. [M.R., 275.]

12 to 15 rounds ⎫ the eyesight, training, and physical con-
a minute, but ⎬ dition of the man; visibility of the
varies with :— ⎭ aiming mark; the range. [M.R., 274, 273.]

Only used when the effect obtained in a given time with deliberate fire would not be sufficient. [I.T., 116 (12), 118 (9), 137 (4).]

Always used in short bursts.

Snap shooting.—Used at short range at an enemy only exposed for a few seconds, or when the firer can only expose himself momentarily.

VOLUME OF FIRE

is the amount of fire aimed at a target in a given time. Volume may be increased by: increasing the rate of fire; increasing the number of firers. [I.T., 90 (3).]

Number of firers limited to one per yard. [I.T., 128 (3).]

Practical value of rapid fire: If opposing forces on equal front, the side using greatest volume with accuracy will win.

OBLIQUE AND ENFILADE FIRE.

More effective than frontal fire; usually a surprise. Larger target surface obtained. Against a long line, errors in ranging not so important if enfilade fire used. [I.T., 116 (11), 118 (9).]

COVERING FIRE.

1. Specially arranged covering fire.
2. Mutual support.

1. The fire of artillery, machine guns and infantry specially detailed to assist the advance of the firing line, and to assist in obtaining superiority of fire. [I.T., 121 (8), 118 (6).]

2. The fire and the constant pressing forward of neighbouring units in the firing line. [I.T., 123 (6).]

Overhead fire.—Infantry should not fire over the heads of friendly troops unless they are absolutely safe from their fire (defiladed). [I.T., 121 (8).]

FIRE SHOULD ALWAYS OPEN IN FULL VOLUME. UNSTEADY FIRING ONLY ENCOURAGES THE ENEMY. FIRE MUST BE KEPT UP TILL ITS PURPOSE IS FULFILLED. [M.R., 523; I.T., 116 (14).]

FIRE ACTION—II.

DUTIES OF LEADERS WITH REGARD TO FIRE.

FIRE ORGANISATION.

One of the duties of the higher commanders after reconnaissance is to arrange for the co-operation of artillery, machine guns, and specially detailed bodies of infantry to assist the firing line.

POSITION OF SUBORDINATE COMMANDERS.

Where they can best supervise their commands; watch the enemy; and receive and transmit orders and information. [I.T., 92, 115 (1), (2), 119, 123 (8), 120 (3).]

Range-takers and observers assist subordinate commanders, distributed as required to : Assist in judging distance; observe fire; watch enemy and neighbouring troops; communicate throughout company. (Use rifles when special duties finished.) [I.T., 116 (15).]

FIRE DIRECTION.

Sometimes subject to orders for fire organisation.

Fire direction includes all the duties of company officers, which are necessary to enable the fire-unit commanders under them to handle the fire of their units to the best tactical advantage. [M.R., 270 ; I.T., 116 (2), 123 (9).]

Some duties of company officers in fire direction :—

1. Reconnaissance.
2. Allot front or objective. [I.T., 123 (3).]
3. Carry out " fire organisation " and other orders from superior commanders. [I.T., 116 (6).]
4. Failing orders, to act on their own judgment with regard to opening or withholding fire.
5. Use formations and ground to assist fire effect. [I.T., 118 ; M.R., 185.]
6. Arrange communication with neighbouring troops and with commanding officer. [I.T., 119.]
7. Regulate movement.
8. Select targets to suit requirements of situation.
9. Decide whether fire is to be concentrated or distributed. [M.R., 271.]
10. Decide whether one elevation or combined sights are to be used. [M.R., 192–196 ; I.T., 116 (16).]

11. Consider and regulate the supply of ammunition. [I.T., 123 (3), (7), 121 (11), 166.]
12. Decide on volume required to produce necessary effect in time available. [I.T., 116 (12).]
13. Decide on allowance to be made for side winds. [M.R., 638 (A).]
14. Consider the means available for ranging. [M.R., 301.]
15. Arrange for simultaneous opening of fire by the fire-unit commanders. [M.R., 523.]
16. Give the necessary "Fire Direction" orders.

The importance of the above duties is sufficiently evident to show that fire must be "directed" as long as possible, so that it may be used according to the wishes of the senior company officer on the spot. [I.T., 123 (9).]

FIRE CONTROL.

Fire control includes all the duties of fire-unit commanders in handling the fire of their units according to "Fire Direction" orders. [M.R., 270 ; I.T., 116 (2), 123 (10).]
The normal "fire unit" of infantry is the section, but sometimes it is one or more platoons. [I.T., 116 (3), 6 (4).]
Some duties of fire-unit commanders in "Fire Control":—

1. Observation.
2. Carry out "Fire Direction" orders.
3. Give "Fire Control" orders to the men.
4. If "Fire Direction" orders have not been received, or are incomplete, they must act on their own judgment in deciding all necessary points. [I.T., 116 (2).]
5. Supervise the men of their sections.

The necessity for controlling fire as long as possible has been discussed in "Fire Action—I," but it is obvious that leaders, however junior, must with their glasses be better able to select targets, and know better than private soldiers how to make use of fire under the various conditions of war. [I.T., 123 (12).]

FIRE ORDERS.

"*Fire Organisation*" *Orders* are issued by a commander to secure co-operation in the fire of various arms and units.

"*Fire Direction*" *Orders* given by an Officer or N.C.O. commanding more than one fire unit to their fire-unit commanders. [I.T., 123 (9).] Contain directions as to how the fire of units is to be used. [M.R., 281.]

Written, signalled, passed, or verbal. If passed must comply with I.T., 96 (3). [I.T., 119 (4).]

"*Fire Control*" *Orders* given by fire-unit commanders to their men, passed or verbal. If passed, I.T., 96 (3) must be complied with. [I.T., 123 (10).]

"*Fire Orders*" generally include :—
(In peace time, orders to load.)

CONCENTRATED FIRE.

RANGE to be used.

Generally given first, as :—

(1) Once sights are adjusted, the men can concentrate their whole attention on recognizing the target from which they need not then look away.

(2) Knowledge of the range limits the area to be searched for the target.

Should it, however, be wished to take the mean of distances judged, to save repetition the sighting elevation may be given after the indication.

INDICATION. [M.R., 277.]

The point of aim must be given; this may be part of the actual target or an auxiliary aiming point, or a distance indicated from either.

When no special part of the target or auxiliary aiming point is mentioned, the centre of the lowest visible part is intended.

When an auxiliary aiming point is given, the target should seldom be mentioned as well (except as follows—"Second gun from the right—a small bush on the left of it—AT THAT BUSH.")

NUMBER OF ROUNDS. Normally five, so as to ensure a lull in which fresh orders can be heard. [M.R., 273.]

FIRE OR RAPID FIRE. These are Fire Control Orders.

PARTIAL DISTRIBUTION.

RANGE.

INDICATION (points of aim of units) or (limits between which fire of units is to be applied).

"By Sections" or "by Platoons."

NUMBER OF ROUNDS.

RATE (if rapid only).

SIGNAL to unit commanders to commence.   [M.R., 523, 283.]

These are Fire Direction Orders.

RANGE.

INDICATION (point of aim).

ROUNDS.

"FIRE" or "Rapid Fire."

These are Fire Control Orders resulting from the above Fire Direction Orders.

MAXIMUM DISTRIBUTION.

RANGE.

INDICATION (limits between which fire is to be distributed).

ROUNDS.

"DISTRIBUTE." (To indicate that each man selects a separate aiming point within the allotted limits and approximately opposite to his own position in the fire unit).

"FIRE" or "Rapid Fire."

These are Fire Control Orders.

How Given.   [I.T., 11, 119 (4).]

CALMLY; otherwise confusion.

WITH DECISION; as orders, to command attention and obedience.

LOUD; for everyone concerned to hear.

PAUSES; to allow each part to be understood and acted on (or if necessary repeated).

Every word must be important.   [M.R., 281.]

Avoid conversation; unnecessary or confusing detail; repetition of point of aim or sighting elevation. If no change is to be made after a pause (*e.g.,* 5 ROUNDS), "FIRE" may be sufficient.

Sighting best changed by UP or DOWN, 200, &c.   [M.R., 283.]

Cease fire means reload and wait for orders. The whistle may be used to draw attention.   [M.R., 284.]

Mutual understanding between commanders and their men simplify fire orders.   [M.R., 282.]

Brief orders may be necessary at obvious targets, *e.g.,* at cavalry surprising a line—

"200, $\frac{1}{2}$ left, Rapid Fire."

Anticipatory orders, e.g., anticipating a movement of the enemy. [M.R., 281, 282, 292.]

"When they reach the wire fence—5 rounds—Rapid Fire."

In order to facilitate control outside decisive range, it is to be understood that in cases where the number of rounds is not given in a "fire control" order, the men should be trained, when reloading, to look to their section commander or nearest fire-unit commander, in case further orders are to be issued,

METHOD OF PRACTISING FIRE ORDERS.

Fire orders should be frequently written down and afterwards discussed.

Communicating instruction.—Fire-unit commanders are formed up not less than fifty yards apart. The instructor gives a fire order to one, which is repeated by each in turn; the instructor checks the accuracy of the order as it is passed, and criticises the way in which it is given.

Control on men.—Concealed fatiguemen are called up individually and fire blank—meanwhile the squad (of N.C.O's.) except the commander is turned away. The fatiguemen again take cover; the squad is turned about—the commander gives his fire orders—the squad adjust sights and lay rifles from rests on the point at which they would have fired. The fatigueman is again called up, aims and sights checked, and the distance taken with a range-finder. The fire orders and probable effect of fire are then criticised. As progress is made two fatiguemen may be called up at a time and orders given for distribution between the points which they mark.

Use of key rifles.—These may be laid on points instead of using fatiguemen; when using landscape targets pins may be used to show the points on the miniature.

Control on dummy screens.—Suitable for a squad or section. Fatiguemen with dummy screens may be concealed and called up when required. [M.R., Part II, Plate 31.]

They may be used to represent small bodies of infantry in various formations, according to the number of dummy screens used and their arrangement, *e.g.*—

A small body advancing in close formation.
A small body moving to a flank.
An extended line firing, or advancing.
A firing line being reinforced.

Other targets may be used either separately or together with dummy screens. [M.R., Part II, Plate 30.]

Note.—In all the above exercises, unless the targets obviously represent service targets, commanders should be told what they are supposed to engage, *e.g.*, in control on men, when a single man appears a commander might be told that a machine-gun is concealed at the point marked by the fatigueman. Only such fire as the supposed target justifies should be allowed.

Dummy screen exercises.—Suitable for one or more platoons. These are not tactical exercises, but are framed in order to practise commanders in giving fire direction orders, and fire-unit commanders in giving fire control orders. The suitability or otherwise of the orders should be discussed with reference to the nature of the targets, but without reference to possible tactical results. At first, direction orders must be given, and these must be accurately carried out by fire-unit commanders, the orders in each case being full and detailed under a variety of situations. As proficiency increases, direction orders should be curtailed to develop initiative and judgment in fire-unit commanders. The final object to be aimed at is the reduction of orders to a minimum without loss of control or effect.

EXERCISES IN FIRE DIRECTION AND CONTROL.
[M.R., 537, 353.]

The object of these exercises is to train officers and N.C.O's. of companies in working out, on the ground, without troops, such problems dealing with fire direction and control as they might be expected to meet in war. The force whose action is considered in detail should seldom consist of more than a company. The force must always be a portion of a large force of all arms and the relation of the small force towards the larger one of which it forms part, and the fire action of all arms must be kept in view throughout the exercise. [Train. and Man. Regs., 20, 38 (3); I.T., 107 (9).]

Examples.

1. *Defence* (active).

1st Phase.—The general position to be occupied, by the force is selected beforehand. The sector for which the force is responsible is pointed out.

Points for consideration :—

    (*a*) Clearance of foreground.
    (*b*) Siting of trenches.
    (*c*) Preparation of range-cards.
    (*d*) Types of trench suitable.
    (*e*) Concealment of trenches (remembering aircraft).
    (*f*) Obstacles.
    (*g*) Communication trenches, &c.

2nd Phase.—Advanced troops considered to have fallen back :— Dummy screens or other targets can be made to appear.

Points for consideration :—

    (*a*) Location of service targets; use of field glasses.
    (*b*) Selection of targets.
    (*c*) Ranging (judging distance combined with use of range-cards).
    (*d*) Fire orders (these should be written).

2. *Attack* (decisive).

Position of enemy's infantry and artillery pointed out ; position of own advanced troops shown ; object of force detailed.

1st Phase.—Points for consideration :—

    (*a*) Use of lateral judging distance.
    (*b*) Preparation of range-cards.
    (*c*) Preliminary reconnaissance (covered approaches, probable fire positions, range marks of enemy, obstacles).

(*d*) The nature of the enemy's fire to be expected.
(*e*) Formations to be first used (the actual line of advance of one of them should be followed).
(*f*) The probable necessity for changes of formation during the advance.

2nd Phase.—The position of advanced troops having been reached :—

(*a*) Question of opening fire : orders for.
(*b*) Method of further advance.
(*c*) Reconnaissance from service position before advance.
(*d*) A fire-unit commander actually advances to selected position.
(*e*) Criticism of position selected.
(*f*) Use of range-card during advance.
(*g*) Question of supporting other parts of line.

FIRE CONTROL EXERCISE WITH MEN.

(Suitable as a Platoon or Company Exercise.)

An instructional exercise, with no tactical consideration, for the teaching of Fire Orders, Fire Control, Passing of Orders, and Fire Discipline.

It can be carried out either in open country or on landscape targets. The former is, of course, best.

Open Country.

1st Phase.

1. Select a suitable piece of ground, giving good landscape in front.

2. Make up good fire orders ; if possible, with assistance of one or two others. Each order should be framed to meet some imaginary situation.

3. Take ranges, or estimate distance as accurately as possible, to the various targets selected.

4. Choose exact position for each squad, so as to ensure that each target is visible to all.

5. Place squads in position, in line, about 10 paces between squads ; the fire-unit commander . kneeling behind his squad. As proficiency increases the squads can be placed further apart.

6. Give out the orders quietly to one commander, and have them passed from commander to commander along the line.

7. As a squad commander shouts the orders to the next commander, the men of his squad act on the orders.

8 After the orders have been given, commanders note how many men fired at correct target and other points as in fire discipline training.

9. General criticism by the directing officer.

2nd Phase.

1. Commanders are called together and told that a certain situation exists.

2. They then return to their squads and give fire orders to meet that situation.

3. Either :—

> (a) Each commander writes down his orders before giving them, or
>
> (b) Each commander has behind him a man who writes down what the commander actually says.

4. This enables all the orders given to be afterwards criticised.

5. The original orders having been made up with a view to some definite situation existing, these situations can then be used in the second stage.

Landscape Targets.

1. The same general arrangements apply as for open country, but each squad must have a landscape target placed in front of it.

2. The landscape targets must all be the same picture.

3. Carry out as for open country, first and second stage.

FIRE DIRECTION PRACTICES.

(Part V, Table B.)

In the various "Methods of Practising Fire Orders' and in "Exercises in Fire Direction and Control," company officers and N.C.O's. have studied and practised all the essentials of fire direction and control; but certain details have been purely theoretical, and can only be checked by the actual use of service ammunition; this is necessary before the commencement of collective field practices. [M.R., 534.]

These details are :—

(1) Judging the effect of atmosphere on the elevation required.

(2) Judging the effect of wind on the flight of the bullet.

(3) Verifying sighting by trial shots.

(4) Observation of fire.

In Part V, Table B, practical demonstration of these points is given. About 300 rounds per battalion should be sufficient. [M.R., 535, 539.]

Practices are fired by parties of officers and N.C.O's. Not necessary for the same practice to be fired by every company. Better to vary the practices, and for all officers and N.C.O's. of battalion to be present and study the results. [M.R., 536.]

Every officer and N.C.O. should write his orders for fire direction before firing takes place; these should be afterwards criticised on the results obtained. [M.R., 538.]

First, fire should be opened at a distinct target at unknown, but within effective, range, and the fire applied to the target from observation. [M.R., 538.]

Markers should be used to confirm the observation. [M.R., 538.]

Signals used by markers. [I.T., 164.]

On range being obtained, the sighting elevation used should be compared with distance as measured from maps and range-finders. Other practices should be framed to prove the effect of wind. The results should be applied at long range, the distance to the target having been given as found with a range-finder or map. Wind gauge may be used to more accurately measure the effect of wind. [M.R., 538, 321.]

NOTE AS TO MUSKETRY TRAINING IN FRANCE.

In inclement weather, when other training is difficult, it is always possible to carry out the Musketry Standard tests.

Ranges, tripods, aim correctors, &c., can be easily improvised with a little ingenuity in places ill provided with range accommodation, and the absence of a supply of the normal equipment is no excuse for the cessation of musketry instruction.

Rifle meetings and competitions with unusual events, such as "gas attack," &c., do much to stimulate and maintain the general musketry proficiency.

www.ingramcontent.com/pod-product-compliance
Lightning Source LLC
Chambersburg PA
CBHW032217040426

42449CB00005B/644